WHAT'S IN THE SKY?

STARS AND PLANETS

by Thomas K. Adamson

a Capstone company — publishers for children

Raintree is an imprint of Capstone Global Library Limited, a company incorporated in England and Wales having its registered office at 264 Banbury Road, Oxford, OX2 7DY – Registered company number: 6695582

www.raintree.co.uk
myorders@raintree.co.uk

Hardback edition © Capstone Global Library Limited 2023
Paperback edition © Capstone Global Library Limited 2024
The moral rights of the proprietor have been asserted.

All rights reserved. No part of this publication may be reproduced in any form or by any means (including photocopying or storing it in any medium by electronic means and whether or not transiently or incidentally to some other use of this publication) without the written permission of the copyright owner, except in accordance with the provisions of the Copyright, Designs and Patents Act 1988 or under the terms of a licence issued by the Copyright Licensing Agency, 5th Floor, Shackleton House, 4 Battle Bridge Lane, London SE1 2HX (www.cla.co.uk). Applications for the copyright owner's written permission should be addressed to the publisher.

Edited by Alison Deering
Designed by Sarah Bennett
Original illustrations © Capstone Global Library Limited 2023
Picture research by Julie De Adder and Svetlana Zhurkin
Production by Katy LaVigne
Originated by Capstone Global Library Ltd

978 1 3982 4795 6 (hardback)
978 1 3982 4799 4 (paperback)

British Library Cataloguing in Publication Data
A full catalogue record for this book is available from the British Library.

Acknowledgements
We would like to thank the following for permission to reproduce photographs: Shutterstock: Aphelleon, 4 (top left) and throughout, Dean Drobot, 21, Dima Zel, 10, elladoro, 7, imagedb, 20, Nowwy Jirawat, 1, Nurhuda Rahmadihan, 11, Ralf Juergen Kraft, 17, Sam Wagner, 16, sripfoto, 12–13, Standret, cover, Tanya Zima, 19, Viktor Malyshchyts, 15, w.aoki, 4–5, Yuriy Kulik, 6, 8

Every effort has been made to contact copyright holders of material reproduced in this book. Any omissions will be rectified in subsequent printings if notice is given to the publisher.

All the internet addresses (URLs) given in this book were valid at the time of going to press. However, due to the dynamic nature of the internet, some addresses may have changed, or sites may have changed or ceased to exist since publication. While the author and publisher regret any inconvenience this may cause readers, no responsibility for any such changes can be accepted by either the author or the publisher.

Printed and bound in India.

Contents

What are stars?..4

What are constellations?..6

Do stars move?..8

Do we see different stars from different parts of Earth?..........10

Why do some stars look brighter?..12

What are planets?..14

Do planets move?..16

Which planets can we see from Earth?......................................18

    Star light, star bright..20

    Glossary...22

    Find out more...23

    Index..24

    About the author...24

Words in **bold** are in the glossary.

What are stars?

Look up at the sky on a clear night. Most of the dots of light you can see are **stars**. Stars are huge, glowing balls of bright, hot **gases**. They make their own heat and light.

We usually only see stars at night. They are in the sky during the day too. The Sun makes the sky too bright for us to see them.

What are constellations?

Ursa Major constellation

Groups of stars appear to form patterns. These are called **constellations**. They can look like objects, animals or people. There are 88 official constellations.

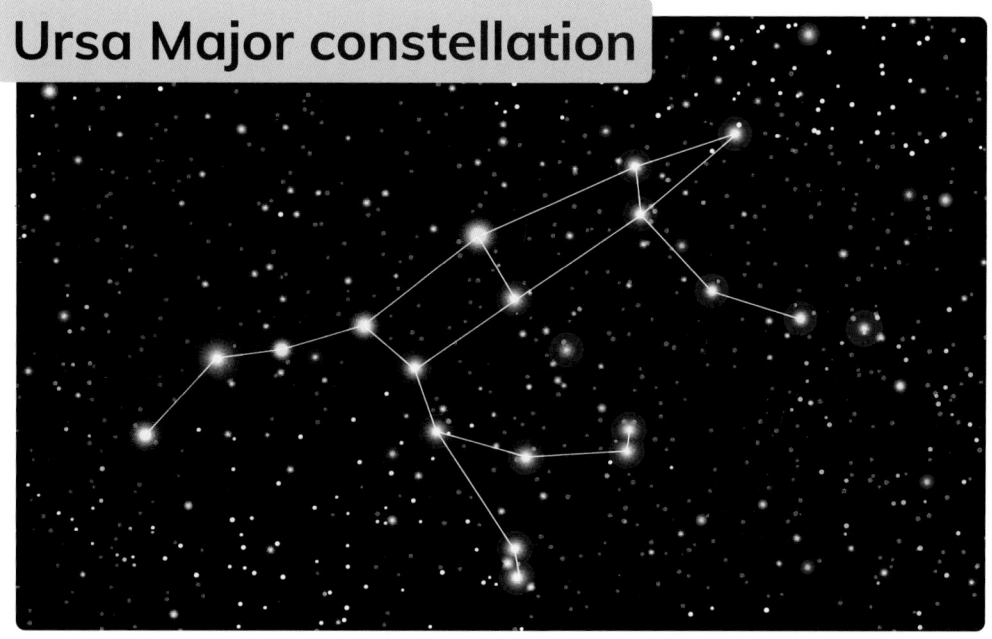

Ursa Major constellation

Some stars in the same constellation might be close together. But most are very far away from each other.

Do stars move?

Orion constellation

Find a constellation in the sky. Look again a few hours later. Those stars will be in a different part of the sky. The stars aren't moving. We are!

Earth spins. That makes it seem like stars are moving across the sky. Earth moves around the Sun too. That makes constellations move throughout the year.

Do we see different stars from different parts of Earth?

Yes! There is an imaginary line around Earth, called the **equator**. It divides our **planet** into two **hemispheres**.

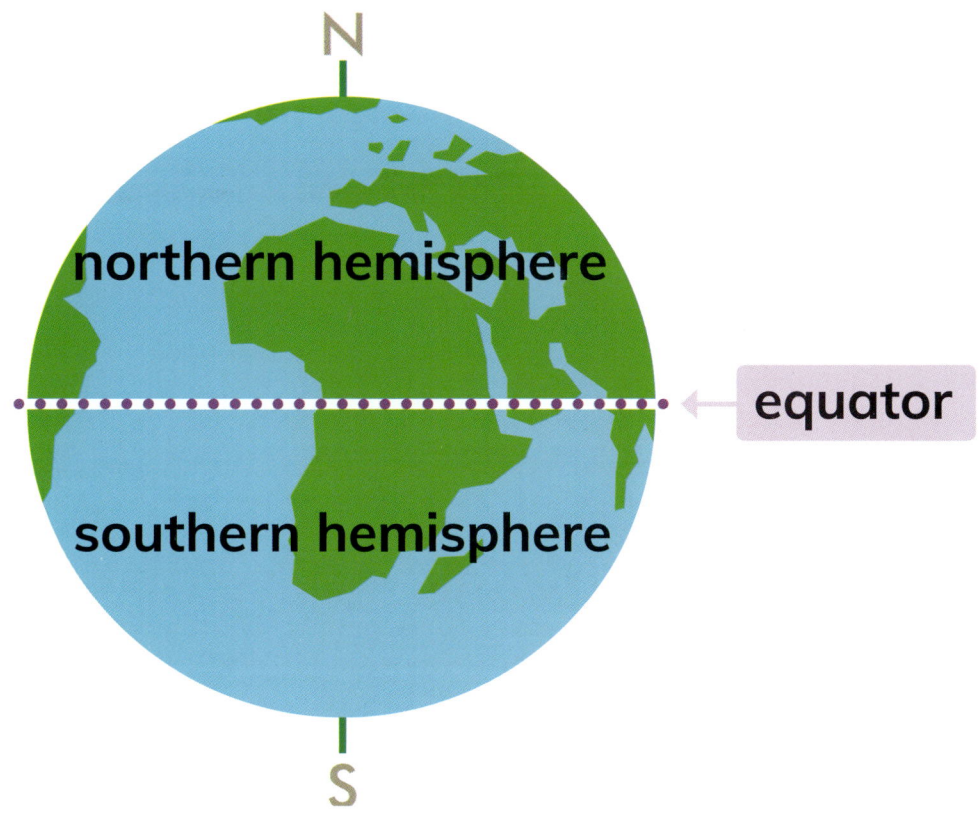

What we see in the sky depends on how far north or south of the equator we are. People in different hemispheres see some different stars and constellations.

Why do some stars look brighter?

12

Not all stars are the same! Some are much further away from Earth than others. Some are even much larger than the Sun.

Stars are also different **temperatures**. Some are very hot. That makes them shine brighter. A star much further away might look brighter because it is hotter than a star that is closer to us.

What are planets?

Planets are large, round objects that move around the Sun. They might look like stars, but they are actually much smaller.

Planets do not give off their own light. They shine because light from the Sun **reflects** off them. They look bright because they are much closer to Earth than stars are.

Do planets move?

Watch the same part of the sky at the same time every night. Stars seem to be in the same place. But planets seem to move from night to night.

Jupiter

Planets **revolve** around the Sun. Earth also spins. These movements make planets appear in different parts of the sky.

Which planets can we see from Earth?

Our **solar system** is made up of the Sun and everything that moves around it. That includes Earth and seven other planets.

From Earth, we can see Mercury, Venus, Mars, Jupiter and Saturn. The other planets are Uranus and Neptune. They are too far from Earth to see without a **telescope**.

Star light, star bright

We need darkness to see the stars in the sky. Try this activity to understand why.

What you need

- gym (or another large room)
- 2 torches
- 2 other people to help

What to do

1. Stand in one corner of the room with all the lights on. Get one person to stand close to you. The other person should stand in the furthest corner.

2. Ask both people to turn on their torches and point them towards you. They shouldn't shine in your eyes, just in your direction.

3. Study the torches. Do they look bright or dim? Does one light look brighter?

4. Now try turning off all the lights in the room. Repeat steps 2 and 3. Which light looks brighter now?

Just like with the stars, the lights are easier to see in the dark. (With all the lights on, it might be hard to tell if the torches are on at all!) But even the same amount of light will look much brighter if it's closer to you.

Glossary

constellation group of stars that forms a shape

equator imaginary line around the middle of Earth

gas substance with no fixed shape that expands to fill any space that holds it

hemisphere one half of Earth; the equator divides Earth into northern and southern hemispheres

planet large object that moves around a star; Earth is a planet

reflect return light from an object

revolve turn or circle around another object

solar system Sun and the objects that move around it

star large ball of burning gases in space

telescope tool people use to look at objects in space; telescopes make objects in space look closer than they really are

temperature how hot or cold something is

Find out more

Books

Big Book of Stars and Planets, Emily Bone (Usborne Publishing, 2016)

Earth and Other Planets (Our Place in the Universe), Ellen Labrecque (Raintree, 2020)

The Solar System (Usborne Beginners), Emily Bone (Usborne Publishing, 2010)

Websites

www.bbc.co.uk/bitesize/topics/zjmqkmn/articles/ztsqj6f
Learn more about the solar system.

www.dkfindout.com/uk/space
Find out more about space.

Index

brightness 13, 14
constellations 7, 9, 11
daytime 4
distance 13, 14
Earth 10–11, 13, 14, 17, 18
equator 10–11
gases 4
hemispheres 10–11
Jupiter 18
light 4, 14
Mars 18
Mercury 18
movement 9, 16–17
Neptune 18
night-time 4, 16
Saturn 18
size 13, 14
solar system 18–19
Sun 4, 9, 13, 14, 18
telescopes 18
temperature 4, 13
Uranus 18
Venus 18

About the author

Thomas K. Adamson has written lots of non-fiction books for kids. Sport, maths, science, cool vehicles – a little of everything! When not writing, he likes to hike, watch films, eat pizza and, of course, read. Tom lives in South Dakota, USA, with his wife, two sons and a Morkie called Moe.